The P–08 Parabellum

LUGER

Automatic Pistol

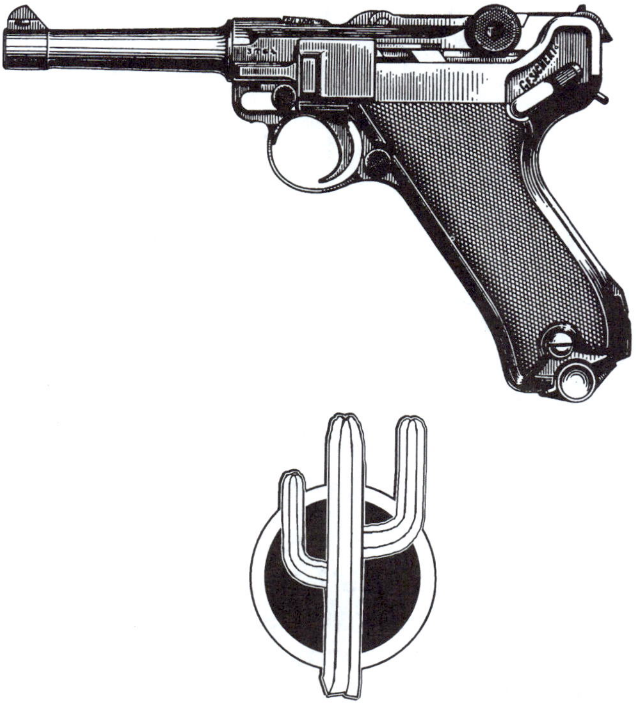

Desert Publications

El Dorado, AR 71731-1751 U. S. A.

The P-08 Parabellum Luger Automatic Pistol

215 S. Washington Ave.
El Dorado, AR 71730 U.S.A.
1-870-862-3811
info@deltapress.com

Printed in U.S.A
ISBN: 978-0-87947-126-2
12 11 10 9 8 7

Desert Publication is a division of
The DELTA GROUP, Ltd.

Direct all inquiries & orders to the above address.

Neither the author nor the publisher assumes any responsibility for the use or misuse of the information contained in this book. This material was compiled for educational and entertainment purposes and one should not construe that any other purpose is suggested.

Warning!

The Publisher (Desert Publications) produces this book for informational and entertainment purposes and under no circumstances advises, encourages or approves of use of this material in any manner.

TABLE OF CONTENTS

* *Indented Chapters Have Been Adapted From The Original DWM Pre-World War I Manual.*

Historical Background

Luger — the very word conjures up visions of a stalemate in the trenches in 1917, or a guerrilla action with captured weapons by French Resistance fighters in Fortress Europe, or a silenced "hit" in the alleys of Istanbul, Port Said or Budapest as Communism and democracy locked horns in the Cold War which followed the fall of the Third Reich. The Luger has a mystique unequalled by that of any other handgun, even the legendary Colt Single Action Army or the Broomhandle Mauser. Some shooters almost become glassy eyed at the mere mention of the Luger, their reactions and emotions reaching an even higher pitch when they get to touch one. And if they actually get to *fire* it— WOW!

Like many legends, all the legendary things that are accredited to the Luger are really just "the best part of the truth", with the gun's failings passed over or swept into the bottom of the wooden shoulder stock holster. Our desire here is to neither increase the legend nor defame this famous weapon, but to present a brief, objective history of the gun and its service record.

Although virtually everyone is aware of the fact that the Luger is a distinctly German weapon, very few people realize that its origins can be traced indirectly back to the Winchester plant in the United States. Putting the experience he had gained with Winchester to good advantage, a former employee by the name of Hugo Borchardt developed a rather strange looking but nevertheless practical semi-automatic pistol. Unable to interest any American firms in manufacturing the gun, he followed the course of John M. Browning and took his design to Europe. He found employment as an engineer at the Berlin plant of Ludwig Loewe and Company. That firm introduced Borchardt's pistol in 1893 and it has the distinction of being the first commercially successful autoloading pistol in the world. It was chambered for the long obsolete 7.63mm Borchardt cartridge which is very similar to the 7.63 Mauser that is still fairly common in Europe and Asia.

Although the success of the Borchardt was short-lived, it did, nevertheless, set the stage for one of Borchardt's assistants to invent a better mousetrap. That young man's name was Georg Luger.

Unlike firearms inventors and designers such as John M. Browning, Hiram S. Maxim, Samuel Colt and others with a number of designs to their credit, Luger is noted solely for designing the one pistol that bore his name. But that association is, of course, one of which anyone would be justifiably proud. It would probably be unfair to say that Luger copied Borchardt's design, especially in the light of the superiority of the later gun. It would probably be more accurate to say that he expanded upon the Borchardt design, or used it merely as inspiration.

The Luger pistol was manufactured by Deutsch Waffen- und Munitionsfabriken (DWM) of Berlin, first appearing on the market in 1900. It was in caliber 7.65mm Parabellum. This new cartridge was a high velocity, bottleneck design which is still manufactured. It is known as the .30 Luger in the United States. The basic mechanical functionng of virtually all Lugers will be covered in a later chapter, but it should be noted here that the earlier models were equipped with grip safeties and toggle locks, and had a leaf-type recoil spring which was subject to breakage. The grip safety and toggle lock were dropped from later production model guns and the leaf recoil spring was replaced with the more rugged coil spring.

P-08 PARABELLUM

When the German military showed interest in the Luger pistol, but suggested a more powerful cartridge, the 7.65mm case was simply modified from a bottleneck to a straight walled type to create the most popular semi-auto cartridge in the world — the 9mm Parabellum. Since the word ' parabellum" translates into English as "prepare for war", it is only fitting that more countries have adopted it for their sidearms and submachine guns than any other cartridge. The 9mm Parabellum appeared in 1902.

That year also marked the appearance of the Model 1902 *Jagd-Karabiner*, or hunting carbine. These had an 11.8" barrel and were supplied with a removable shoulder stock. Although 7.65 Parabellum was the standard caliber for these carbines, at least one specimen is known in 9mm. Legend has it that the carbine was designed for Kaiser Wilhelm, who loved to hunt but was handicapped with a withered left arm, hardly conducive to aiming a rifle. An interesting story if not a verifiable fact. Prototype carbines were made as early as 1900 and it is certain that they were in production by 1904. Production ceased during World War I since the carbine had little military value. It was, however, tested in combat and in that guise it was often fitted with a 32 round snail magazine. Noncommissioned officers in machine gun crews were generally issued these guns, which offered greater range and firepower than the standard sidearm without the bulk and weight of a rifle. The appearance of the submachine gun made such pistol carbines obsolete for military use. Approximately 4,000 were produced before the war, and a handful were assembled, most likely from parts on hand, in the 1920's.

When the U.S. Army announced that it was searching an autoloading pistol in .45 caliber, a few .45 Lugers were made up by hand for the tests. The arm was rejected, and these guns, if they still exist, are extremely valuable collectors' items. In *Death Sentence*, author Brian Garfield has a vigilante in Chicago making "hits" with a Luger in .45 ACP caliber, a thoroughly preposterous situation for several reasons. First, the gun is so rare that the vigilante would not have found it sitting in a gun shop, when it changed hands it

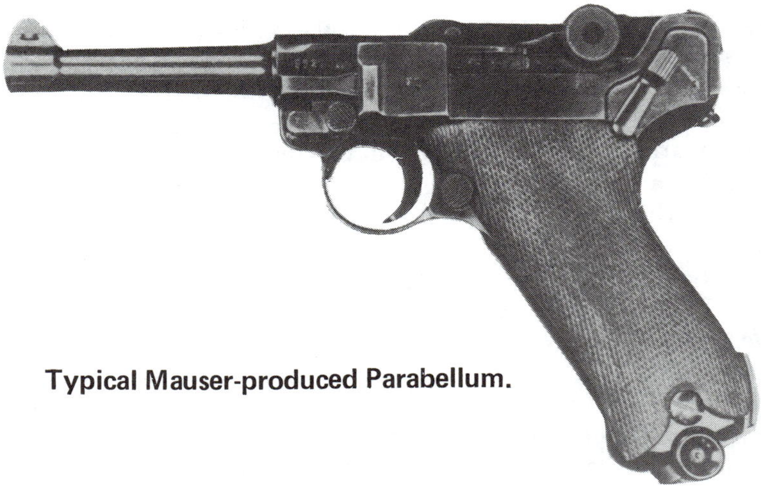

Typical Mauser-produced Parabellum.

would undoubtedly have gone from one collector to another. Second, the gun's price would have been far out of reach of even a well to do person, let alone the average working man the vigilante was depicted as being. Third, no one in his right mind would use such a rare gun to make a "hit", since it would certainly make the police ballistics experts take notice (as they did in the book). Since the author is basically a very competent writer, I can't help but think he inserted the .45 Luger into the story with a tongue in cheek attitude.

The Luger was adopted by the German Navy in 1906, and the Army accepted the P-08 for service in 1908. The Navy Luger had a 6" barrel, adjustable rear sight and shoulder stock while the Army weapon was equipped with a 4" barrel and fixed sights. An 8" barrelled model was also produced and is often referred to as the Artillery Model. It had a tangent rear sight and was furnished with a wooden holster/shoulder stock in the same manner as the Broomhandle Mauser of the same era.

Although rejected by U.S. forces, the Luger was readily accepted by many European and South American countries which lacked the technology and personnel to produce suitable handguns of their own. These countries included Bulgaria, Portugal, Switzerland, Bolivia and Chile among others. Switzerland is, of course, an exception to the rule of the above since Swiss technology, skilled personnel and production quality have always been on a standard with the best in the world.

With the assassination of Archduke Francis Ferdinand and his wife Sophie in June of 1914, the world was plunged into a "war to end all wars". Although autoloading pistols were a fairly recent development, virtually every major power that participated in the conflict had an autoloading service pistol, whether standard issue or substitute standard. Germany was, of course, armed with the Luger and Broomhandle Mauser as her principal sidearms. England had the Webley Mark I autoloader as standard issue to the Navy. France was armed with obsolete 8mm and 11mm revolvers but purchased large quantities of Star and Ruby autoloaders from Spain in 7.65mm (.32 ACP). Italian troops were armed with the Glisenti and Beretta. Austro-Hungary had the Roth-Steyr, Austria the Steyr-Hahn and Hungary the Frommer. (Just to avoid confusion on that last part, Austro-Hungary refers to the two countries together, Austria or Hungary would, of course, refer to only one of them). The United States was armed principally with the Colt Model 1911, supplemented by Smith & Wesson and Colt Model 1917 revolvers. Of all the guns listed above, only the Luger and the Colt are easily recognized by anyone but a firearms expert.

In actual service, the Luger proved to be a poor choice as a military sidearm, although in overall performance it was still superior to most others of the day. Its toggle form of operation meant that a lot of delicate moving parts were exposed to dust, mud, grit and abrasion, hardly a desirable feature on the battlefield. The gun's design was rather complex and expensive to manufacture. The action contained a lot of delicately balanced cams, springs and lever arms. The trigger assembly had a good many moving parts and the trigger pull generally ran from poor to fair. The sights were poor, consisting (on the P—08) of a V notch rear and inverted V front. The action was extremely sensitive to the ammunition used. Despite its poor sights and generally poor trigger pull, the Luger did gain a reputation for accuracy based largely on its natural pointing qualities. And no one could argue that the gun was beautifully made. A magnificent piece of machinery it is, but a serious service weapon it's not, even if it was used by Germany in two World Wars.

Following the defeat of Germany in 1918, the allies disbanded the DWM complex and commercial production of the Luger was moved to the Mauser plant at Oberndorf. Prior to and during World War I, the P-08 was also produced by Erfurt Arsenal, Simson & Co., Vickers Ltd., Waffenfabrik Bern and Heinrich Krieghoff Waffenfabrik.

Despite its weak points as a military arm listed earlier, the P-08 continued to enjoy widespread success both as a commercial and military arm. A number of wartime production guns were reworked and sold to Russia. Finland armed her Army, Navy and Air Force with the Luger. The Lithuanian Border Guards were like armed and a good many Lugers were sold to Turkey and Persia.

The leading American importer was, of course, the famous A. F. Stoeger organization. Lugers made for export to the United States were typically marked with the American Eagle and that symbol is as closely associated with Stoeger as Tarzan is with Jane. In fact, the word "Luger" is a registered trademark of Stoeger and has been since the 1930's. Technically, it is incorrect to call any Parabellum produced since then a Luger, although everyone does it. In contrast, while the current Stoeger .22 Long Rifle Luger isn't really a Luger, it is because they own the name.

When Germany began to rearm in the 1930's, the German high command decided to replace the P-08 with a weapon that was both easier and cheaper to manufacture. The resulting weapon was the Walther P-38. Although possessed of the dubious virtue of being a double action autoloader, the P-38 is nevertheless noteworthy for its reliability. The Walther was adopted in 1938, but the P-08 remained in production until 1943. Some 400,000 Parabellums were produced for the German Army *after* the acceptance of the P-38.

P-08 PARABELLUM

Mauser resumed production of the Parabellum in 1971, offering both 1906 and P-08 models which curiously are fitted with the grip safety of the earliest models. In 1979, 250 Parabellums were made in Model 1902 carbine configuration. These were not true replicas in that they feature the flat toggle grips and coil mainspring of later variations, but they are an interesting collector's item. With the typical expensive machine work and high quality of the Parabellum, the correspondingly high price for post-1971 models kept sales down. Production of the Parabellum, like the Colt Single Action Army, has finally been discontinued, most likely for good. A smaller scale version of the P-08 in .380 ACP was offered in the 1960's, but that little gem didn't last long, either. The only Lugers still in production are the .22 Stoegers mentioned earlier.

Due to its generally excellent workmanship and finish, the variety of markings to be found on both military and commercial models, and its place in history, plus other, less tangible factors which can be understood only by the collectors themselves, the Parabellum is one of the most sought after collectible guns in the world. Except for the post-1971 production run, the only Parabellum that should be considered as a "shooter" is the standard military P-08. All other models are simply too valuable, and the older ones are more subject to breakage due to the inferior metallurgy of yesteryear. For shooting purposes, probably the most desirable P-08s are the Mauser-produced guns from the late 1930's. After nearly 40 years of production, the bugs had long since been worked out of the production process and metallurgy had also taken great strides. However, even these guns should not be abused, but should be given the same care that any decent piece of machinery deserves.

With its serious faults, one might wonder why the Luger attained a fame and mystique peculiar to a rare handful of handguns. However, careful study would also show that the Broomhandle Mauser, Colt M1911 and Colt Single Action Army also had their share of faults. But they undoubtedly all succeeded and achieved their well deserved fame for the simple reason that at the time they were adopted, they were the best that was available. Although each was obsolete within 20 years of introduction, they didn't follow the world standard — *they were the standard.*

Principal Features

The Parabellum autoloading pistol belongs to that class of firearms which are recoil actuated, with the action locked while the bullet travels down the barrel. In this way no gases can escape; they maintain a uniform full effect upon the bullet, a condition which alone assures reliable efficiency which for all purposes is essential, but indispensible in warfare.

The breech mechanism — the most important feature of any autoloader — corresponds in principle with the world famous Maxim gun.

Its peculiarly constructed toggle-joint lever mechanism provides a totally secure safety lock, the prime necessity for any firearm.

Like most modern weapons, the Parabellum is equipped with an internal hammer since it's a well known fact that

(Editor's note: This and the following chapters are adapted from the original, pre-World War I DWM manual, hence some references such as to the superiority of a removable magazine over a fixed magazine with stripper clips, etc., are hardly applicable to today's arms technology. However, they have been retained in the interest of historical value.)

an exposed, thumb-cocked hammer is one of the leading causes of firearm accidents through being caught, etc.

In connection with the breech mechanism stands the safety sear (commonly called a grip safety in the United States) which is so arranged that first it blocks automatically whenever the weapon is not held in the hand and also, like all other safety arrangements, can be fixed or brought out of contact by a separate action.

The construction of the extractor allows the shooter to see and feel whether or not there is a round in the chamber, an important safety advantage.

The loading arrangement of the Parabellum gives it a further advantage over some contemporary designs. The use of a removable magazine has the following advantages over stripper clips: in case of danger the attention is not taken off the adversary when reloading, the pistol can be loaded while on foot or on horseback, with stiff or gloved hands and in every position, plus the fact that when the magazine is inserted into or withdrawn from the magazine well the hand is fully protected from being jammed or otherwise injured by the mechanism.

As regards the calibers selected (7.65mm or .30 Luger with .301" dia. bullet and 9mm Parabellum with .354" dia.), the automatic action of the pistol does not in any way depend on the size of the caliber but on the total energy which the bullet develops upon firing and on the amount of recoil thus obtained. This energy may vary to a considerable extent, but its minimum should be sufficient to disable the enemy, that is, to render further fighting impossible on the part of the opposing soldier who is in full service uniform, protected by all sorts of equipment such as belts, cartridge pouches, buckles, metal buttons, fur coat, breastplate, etc., and perhaps under an additional cover of planks or brush, or behind a closed door.

The smaller caliber has more penetrating ability, the larger more stopping power.

Constituent Parts

The parts of the Parabellum pistol are grouped in two divisions like the Maxim weapons, on which this one is based, one forming the part which moves at firing, consisting of barrel, mechanism and breech (Fig. E, Table I). The other one, with the frame containing all the other apparatus which produces the automatic action, representing the fixed part.

The two divisions are:

a. The Moveable Part:

1. Barrel with front sight (1 I) and dividing receiver (1 II).
2. Breech-block, in connection with

(3.) forward and

(4.) rear link of toggle joint, with coupling link from recoil spring (4 I) and pin (4 II),

(5.) connecting pin between the breech-block and the forward link,

(6.) connecting pin between the forward and rearward link.

7. Hinge between the rear link and the divided receiver.

(8.) Rivet for connecting pin (6).

12. Firing pin.
13. Firing pin spring.
14. Breech-block end piece.
15. Cartridge extractor with spring (15 I) and pin (15 II).
16. Cartridge ejector.
18. Trigger bar with spring-stud (18 I), spring (18 II) and rivet (18 III).
19. Trigger bar spring.

b. **The Fixed Part:**

11. Recoil spring with drawing rod (11 I), bell crank (11 II) and pin (11 III).
17. Stock with sling swivel (17 I) and breech-block catch-link rivet (17 II).
20. Trigger with spring (20 I).
21. Trigger plate.
(22.) Trigger lever.
(23.) Trigger lever pin.
24. Locking bolt.
(25). Locking bolt spring.
26. Breech-block catch-link with spring (26 I).
27. Magazine catch.
28. Magazine catch spring.
29. Automatic safety sear (grip safety).
30. Automatic safety sear spring.
31. Safety catch.
32. Safety catch pin.
33. Grips, of wood.
34. Grip screws.
35. Magazine consisting of the following parts which can be separated:
    a. Sheet metal frame.
    b. Cartridge feeder spring.
    c. Cartridge feeder or carrier.
    d. Cartridge feeder pressure knob.
    e. Cartridge feeder guide knob.
    f. Bottom piece.
    g. Connecting pin.

P-08 PARABELLUM

The parts which appear in brackets as well as those which are bracketed together and those with a Roman numeral (1 I, 1 II, 4 I, 4 II, etc.) should never be detached unless for replacement or expedient repair. This should only be done by a competent gunsmith.

Fig. 1

Manipulation And Action

1. Loading The Magazine.

Grasp the magazine in the left hand, with the aperture upwards and its point to the right. Place the left thumb on the checkered guide knob (35e) and pull it down a distance roughly equalling the diameter of a cartridge; simultaneously (as shown in Fig. 1) insert a cartridge with the right hand into the magazine beneath the overlapping lips of the magazine, taking care not to damage the lips. Pull the guid knob down once more and insert another cartridge. Repeat the process until the magazine is full. Always draw the guide knob down *step by step* so as to allow only sufficient space for one cartridge at a time. In this way the strong magazine spring will not needlessly recoil and the correct position of each cartridge is ensured.

The magazine is emptied by removing the cartridges one by one, each time slightly drawing down the guide knob to allow easy removal of the uppermost cartridge. During this operation, as in all others, the guide knob must not be allowed to spring up freely.

2. Inserting the Magazine Into The Pistol.

Hold the pistol firmly in the right hand, as in firing, while the left hand pushes the magazine upward into the magazine

Fig. 2

well until the magazine catch (27) snaps into its seat (u). This is easily done by giving the bottom of the magazine a slight tap with the left hand.

To remove the magazine, hold the pistol in the right hand and turn it slightly to the left as shown in Fig. 2. Press in the magazine catch (27) with the thumb of the right hand. The magazine can easily be removed with the left hand or, in an emergency, simply be allowed to fall to the ground so that the left hand can insert a fresh magazine immediately.

3. Loading The Chamber.

Hold the pistol firmly by grasping the butt with the right hand, the trigger finger stretched out on the right side of the trigger guard. With the barrel pointed in a safe direction, firmly grasp the two cheeks of the toggle-joint (Fig. 3) with the thumb and first finger of the left hand.

Do not merely hook the finger and thumb into the toggle-joint. Pull upward, then tangentially back until the face of the breech-block is *behind* the uppermost cartridge in the magazine, then release the toggle-joint and let the breech-block spring forward abruptly. If the links of the toggle-joint are not completely extended by this action, gentle hand pressure should accomplish this.

Now the top cartridge has been transferred from the magazine to the chamber, the firing pin is cocked and the weapon is ready for firing (Fig. J, Table II).

Single loading, without using a magazine, is accomplished by grasping the pistol in the right hand as previously described. The left hand draws the toggle-joint completely to the rear,

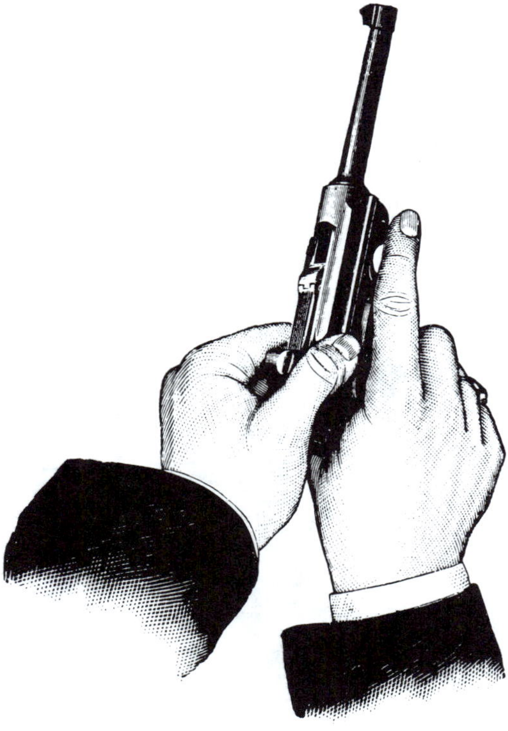

Fig. 3

with the first finger of the left hand holding the left cheek of the toggle-joint *tightly* while the thumb pushes backward against the sling swivel and the middle finger presses against the edge of the trigger plate (Fig. 4), thereby keeping the breech open. With the right hand, insert the cartridge into the chamber. After the cartridge is inserted, close the breech by allowing the toggle-joint to extend in the normal manner. With practice, single loading can be easily done.

If a full magazine is inserted while the chamber is loaded, nine shots can be loaded without interruption.

To unload the weapon, remove the magazine if one is inserted into the weapon. Then carefully work the toggle-joint in such a way as to prevent the cartridge from being ejected to the ground, allowing it to be gently dropped into the open palm of the hand.

4. Firing.

The position of the pistol as shown in Fig. 5 assures a splendid percentage of hits, because the hand is in almost exact prolongation of the axis of the weapon. If held too loosely or too low as shown in Fig. 6, the resistance to recoil is insufficient and the efficiency will be much impaired.

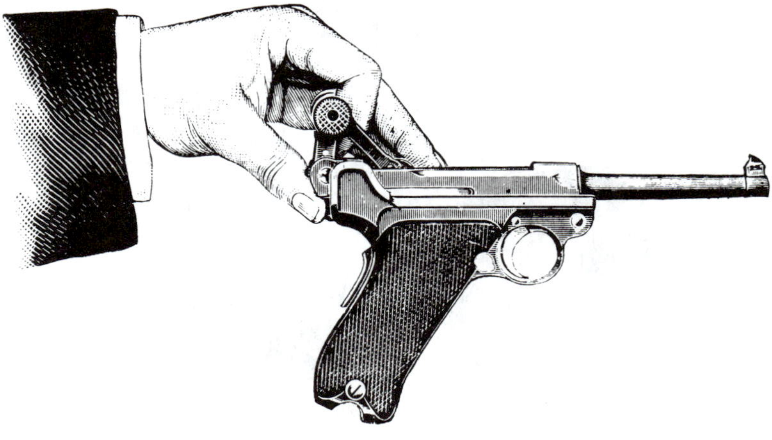

Fig. 4

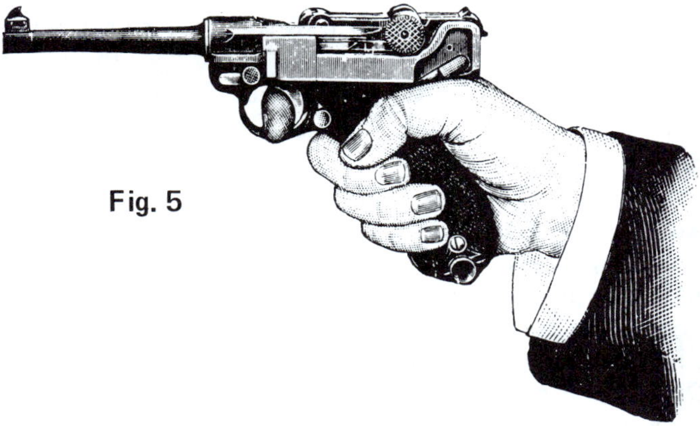

Fig. 5

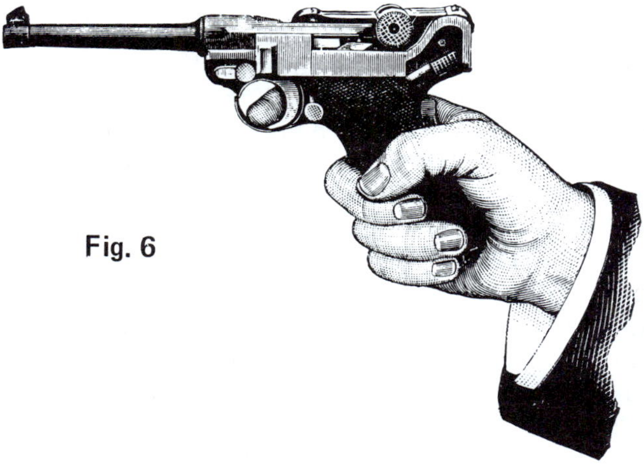

Fig. 6

Therefore, hold the pistol firmly pointed at the target so that the automatic safety sear (29) projecting from the rear of the grip is pressed in. To fire the pistol, pull the trigger gently and release, then repeat as long as the cartridges last. When firing slowly it is advisable to use the touching point of the trigger.

The pressure from the first finger upon the trigger is transferred upon the front portion of the trigger bar, the spring stud (18 I), by means of the lever attached to the plate. This pressure acts upon the lower arm of the trigger lever and by a still further pressure being put upon the trigger the firing pin retaining catch (n I) becomes detached from the trigger bar seat (n II). The firing pin, propelled by its spring, fires the cartridge.

5. Automatic Action Of The Mechanism In Firing.

On firing the pistol the moveable part (Table I, Fig. E) recoils in the grooves of the fixed stock.

The joint remains straight and the breech completely closed while the bullet travels up the bore.

During the continued recoil the link cheeks (c X) meet the corresponding curves (C X) of the side plates at the stock. The connecting pin (6) rises above the connecting pins (5 and 7). While the barrel with the divided receiver (l II) returns only as far as the catch (r) in the grip, the energy during the recoil raises the toggle-joint until the recoil spring (11) is completely compressed and the breech-block (2) reaches its rearmost position (Table II, Fig. L). The spent cartridge, drawn out of the chamber by the extractor (15), strikes against the ejector (16) which intercepts it and is thrown out; the firing pin (12) is by means of the cocking piece (n) of the forward link (3) of the toggle-joint drawn back at the retaining catch (n I) and the firing pin spring is compressed.

The recoil spring (11), cocked but not kept back in the cocked position, propels the breech-block by means of its drawing rod (11 I), the bell crank (11 II) and the coupling link (4 I) which hangs on the rear link (4); this movement is transferred to the barrel and divided receiver which run forward until the catch (r) meets the locking bolt (24); the toggle-joint straightens. Simultaneously the top cartridge is stripped from the magazine by the breech-block and inserted into the chamber, the firing pin retaining catch (n I) is caught by the trigger bar and the firing pin is cocked (Fig. H).

The breech is locked again. In shooting it is secured positively in this position by the middle pin (6) lying a little below the line connecting the outer pins (5 and 7) when the toggle links are straightened. Thus the action of the powder gases presses the middle joint down against the divided receiver (1 II) contrary to the opening direction, at the same time the pin (7) connecting the rear link (4) to the receiver takes up the longitudinal pressure (Fig. J).

After firing, the trigger must be immediately released by pushing the finger forward. This is done so the spring stud (18 I) which during the preceding movement was leaning against the upper arm of the trigger lever (22) can again go forward and effect the next firing. *This is very important.*

The pistol is now automatically reloaded, locked, cocked and ready for firing. Therefore, take care!

After the last shot (i.e., when the magazine is empty), the guide knob (35 e) acts upon the breech-block catch link (26), whose tooth presses into the retaining catch (c) of the breech; the breech remains open, the toggle-joint erect and the line of sight obstructed (Fig. D, Table I).

The recoil spring keeps the joint raised even after the removal of the magazine and leaves the open breech resting on the catch link. This latter is pressed into its seat by its own spring (26 I) as soon as the joint is withdrawn.

After the empty magazine has been removed the breech is left open if it is to be reloaded.

After inserting a new (full) magazine, draw the toggle-joint cheek slightly back with your hand and let it spring forward. The weapon will again be ready for firing.

The breech, if open after the magazine has been emptied or after an empty one has been inserted, cannot be closed until the magazine has been entirely or partially extracted.

The safety sear (grip safety) works in the following manner: The safety lever (29) by action of its spring (30) automatically goes with the upper end (S) upwards over the rear end of the trigger bar (18) and in this position positively prevents both the divided receiver from going back and the firing pin (12) which is always cocked when the pistol is loaded from being tripped. Only by firmly holding the grip and pressing the lower shoulder (S o) against the action of its spring is the upper end (S) disengaged and the moveable parts free to act.

The action of the safety catch (31) is simply that the hook of its lower arm (S II) which extends to the inside is positioned either in front of (Fig. A) or beside (Fig. C) the catch (S I) of the safety lever (29) and thereby prevents or allows movement of the lever.

In case the automatic action is inconvenient for small hands the stud of the safety lever (29) is placed in front of instead of behind the spring (30), between the latter and the intermediate wall of the grip (17). By doing this the safety mechanism does not work automatically. The action can be locked by the safety catch (31) only. In either case the shooter when grasping the grip is shown by the safety shoulder either yielding or remaining stationary whether or not his arm is ready for firing.

Cleaning

With the Parabellum, as with all weapons where smokeless powder is used, the greatest attention must be bestowed on the bore, which should be cleaned as soon as possible after the pistol is used. If there is no time to do this thoroughly at once, then the bore should at least be well greased, for which purpose thick grease must always be ready in the grease box of the cleaning rod. Final cleaning and removal of the grease should be done at the earliest opportunity. To clean, insert an *empty* magazine and open the breech. With the toggle-joint raised, draw a patch which has been saturated in solvent repeatedly through the bore from the muzzle end until all powder residue and dirt have been removed. To prevent rust, coat the bore lightly with oil after thorough cleaning. Any little dirt spots in the breech mechanism may be removed from the top. But if the weapon has seen hard wear and fouling, or malfunctions appear in the mechanism, it should be field stripped, cleaned, lubricated on all working surfaces and proper reassembled.

The magazine as well must be kept clean. To clean, compress the spring as for loading and run a patch which has been soaked in solvent over the interior surfaces. Care should be taken to ensure that all traces of solvent have been wiped from the magazine before inserting fresh cartridges since an excess of solvent can work its way into the cartridge and render the primer and powder inert.

Dismantling And Assembling

a. **For Ordinary Use.**

DISMANTLING

1. Separate the moveable part from the stationary portion.

First remove the magazine and work the action to ensure that the chamber is empty. Then dry fire it once to release the cocked firing pin. Depress the safety sear with the palm of the right hand while the fingers grasp the toggle-joints and pull them back (see Fig. 7). With the left hand, shove the barrel back until the link cheeks touch the cam. Retain it there with the right hand while the left thumb turns the locking bolt lever (24) downward. Let the recoiling portion of the weapon slide forward so that the trigger plate (21) rises and can be lifted off. Next push the barrel and the breech mechanism (i.e., the moveable part of the pistol) forward and remove it from the frame (i.e., the stationary portion).

2. Separate the breech mechanism from the divided receiver.

Withdraw the hinge pin (7) from right to left, after releasing the firing pin by a pressure upon the front part

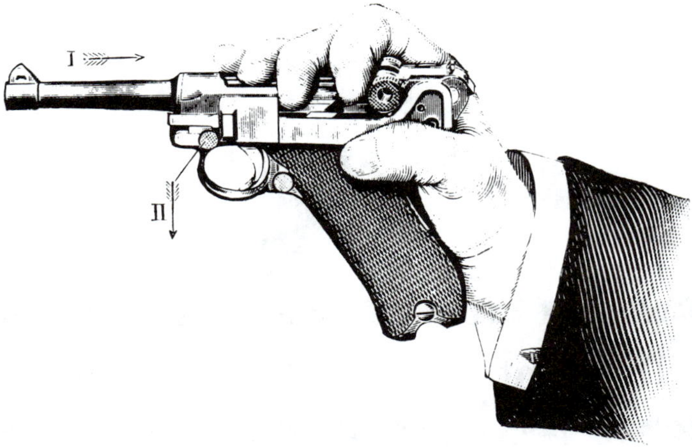

Fig. 7

of the trigger bar (18), then lift the toggle-joint cheeks and withdraw the breech-block.

3. Remove the firing pin from the breech.

Hold the breech tightly (see Fig. 8) with the toggle-joint as straight as possible, use the first finger of the right hand (or a small screwdriver) and press the breech-block end piece (14) firmly inward so as to compress the firing pin spring, turn quickly to the left, thus getting its shoulder out of the groove, and let the bottom pin, yielding to the pressure of the spring, slowly glide out. Remove firing pin and spring.

ASSEMBLING

This is effected in the reverse order of dismantling:

1. Insert the firing pin into the breech-block.

First put firing pin and spring into position, then place the end piece with its shoulder in the slot in the breech-block, and while compressing the spring (the parts can be held and moved as shown in Fig. 8) turn quickly to the right until the shoulder or catch recedes through the transverse

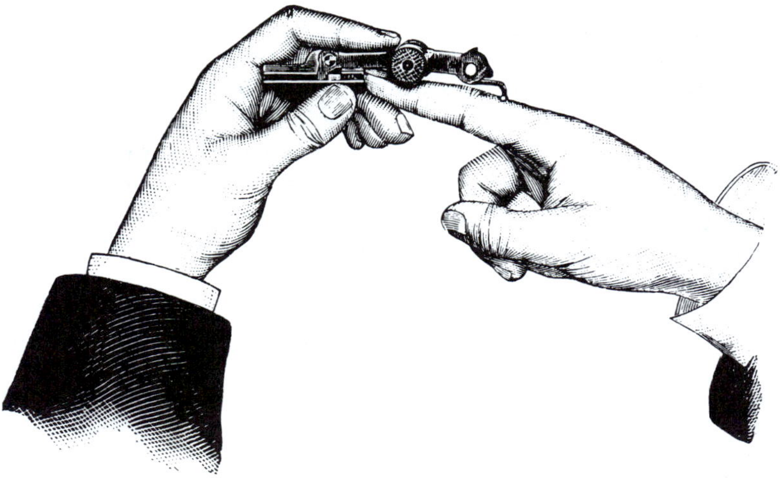

Fig. 8

groove into the axial notch until the line stands *vertically*. It is very important to do this.

2. *Connect breech-block with divided receiver.*

Turn the lower sides of the parts upwards as shown in Fig. 9, lay the guide fillets into the grooves carefully and while pressing down on the front of the trigger bar (in order that the breech-block with the firing pin point can be freely pushed forward) turn the whole assembly around, then insert the hinge pin (7) from the left, thus joining the parts (Fig. E) which move in firing. To ensure proper position for fixing the hinge pin the surfaces (c I, C II) behind the link cheeks must meet correctly.

3. *Join the moveable parts to the stationary portion.*

Hold the former (the barrel with the breech mechanism) in the left hand horizontally, the front sight downward, the coupling link to the rear and *the firing pin always uncocked*. With the right hand, slide the frame carefully onto the receiver (Fig. 10). Turn the pistol right side up, then draw the barrel and the breech mechanism back again until the coupling link (see Fig. 11) can fall into its place in front of the hooks of the recoil spring bell crank with which the breech mechanism engages when pressed back.

There is yet another way of joining the barrel and the breech mechanism to the frame which after some practice is performed more quickly. While barrel and breech mechanism are held in the left hand, the lower side of the parts remain, at first, with the coupling link pointing backward,

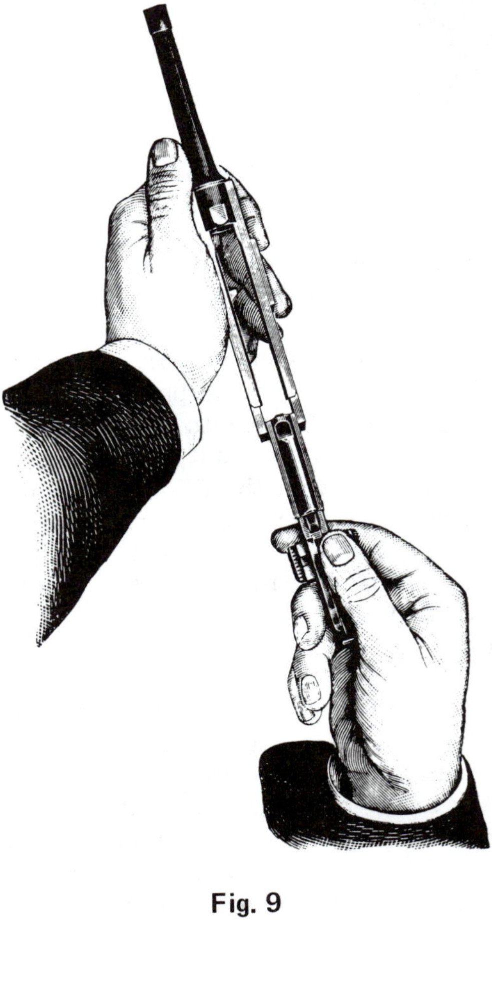

Fig. 9

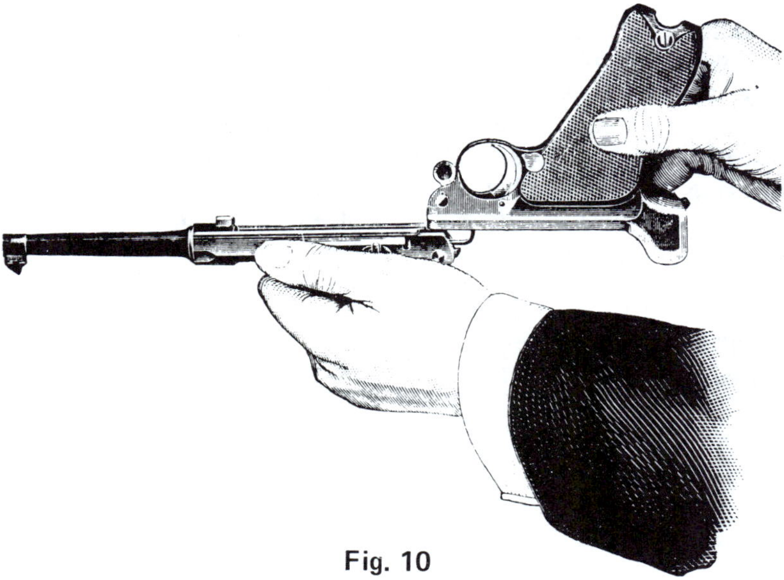

Fig. 10

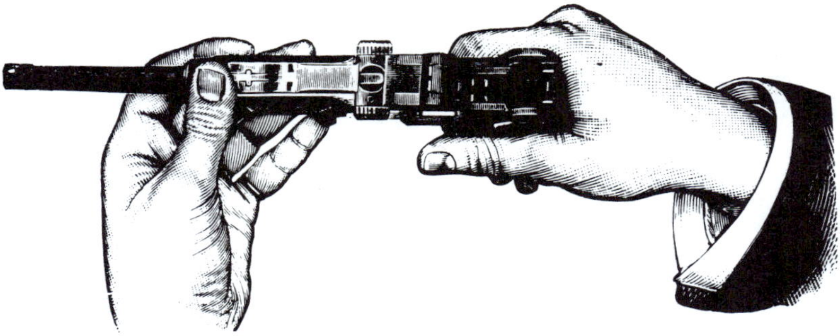

Fig. 11

turned toward the chest. Then the frame, with its lower opening also turned toward the chest, is pushed over the receiver until the coupling link falls in front of the hooks of the recoil spring bell crank as in Fig. 11. Then the whole is held vertically so that when the breech mechanism is pushed in, the link immediately engages with these hooks.

Whether the main parts are joined in the former or the latter way, the pistol is now held as shown in Fig. 7, the safety sear is pressed in and the breech mechanism is pushed back until it reaches the curved surfaces. It is then held tightly in its position, the trigger plate inserted with the rearward projecting fillet beneath the slot in the casing and the small part pointing towards the locking bolt lever (24). This lever is then turned upward.

Now *repeatedly* draw back and release the toggle-joint to make sure the recoil spring is properly caught and works well. *This should be done every time the pistol has been dismantled and reassembled.*

Further Dismantling And Assembling

b. As Far As It Can Be Done Without Special Tools.

Cartridge ejector (16). *To remove:* Barrel with divided receiver is held in the left hand, the thumb nail or a screw-driver lifts the rear end of the ejector *just sufficiently to cause its round shoulder to leave its socket, and lodge against the edge*, the first finger presses from the inside against the catch which projects into the receiver (1 II). The ejector jumps out, to intercept it place the thumb over it.

Another way of removing the ejector is by simultaneously raising the rear end with a thumb nail or small screwdriver and pressing the catch outward with the first finger of the left hand. In neither case should force be used.

(The ejector is only removed to prevent any cotton waste from lodging between the parts when cleaning the barrel from the breech end).

To insert: Fit the part exactly over the corresponding openings of the receiver and press down slightly on the center and rear shoulder. No force should be used.

Trigger bar (18). *To remove:* Slightly lift the trigger bar spring (19) by inserting the thumb nail beneath the upturned part and slide it straight forward. Before doing this, press in the front of the trigger bar so as to lift the spring. Turn the divided receiver to the left and the bar will drop out. If necessary strike the pistol upon the palm of the hand.

To assemble: Insert the trigger bar in its place, the pin pointing forward. Lay the spring into its groove and push straight in while slightly pressing down the center.

Extractor (15). *To remove:* Shove out the pin by means of the punch (III) until the rear part of the extractor is lifted by its spring. Take out the extractor in a forward and the spring in an upward direction.

To insert: First put the spring in place correctly, then from the front push back the extractor along the stroke stops into the breech-block, press it down, fasten it in its place by the pin and make certain it is moving freely. (When shooting frequently this should be done from time to time.)

Breech bolt catch link (26). *To remove:* While gently pressing with the first finger of the right hand or the thumb of the left against the adjacent casing lift it, but no more than it was originally inserted and push straight back.

To insert: Hold the adjusting piece with the thumb and first finger of the right hand so that the middle finger presses the spring from the top. Press the point of the spring down into the slanting notch in the handle and push the link in, in a forward direction while the hook rests underneath the pin. If wrongly inserted, turn the frame upside down and the link will fall out.

Trigger (20). *To remove:* Compress the sprial spring by a slight counter pressure and lift out horizontally.

To insert: Push the spiral spring (with the wire always pointing backward, never sideways) holding it compressed against the slanting part of the casing.

Locking bolt (24). *To remove:* Take hold of it by the lever, give it an upward pressure and pull out in a straight line.

To insert: Lift the lever and press it in.

Automatic safety sear. *To remove:* Unscrew and lift off the left hand stock, press the automatic safety sear (29) against its spring, hold it in its position, lift out the lower end with the stud or catch and pull out the piece downward without using force.

To insert: Push it underneath the safety catch (31) upwards from the left to the right, lift up the stud and drop it into its seat. While this is being done, care must be taken that the shoulder or projection working the spring does not lodge between the spring and the partition in the stock, but stands behind the spring.

Safety catch. *To remove:* With the lower arm turned downward, push out the pin (32) which holds the lever in the frame by means of a punch. Then simply remove the catch.

To insert: Place the part in its right position with the checkered button upward, press the upper arm of the lever and insert completely.

Further dismantling of the Parabellum, especially of the rivetted parts, is almost never necessary and should be left to an expert.

The magazine should not be dismantled expect in special cases. The pin holding the lower part is withdrawn, the spring taken out and the magazine body cleaned from underneath. The feed platform cannot be removed, being rivetted to the cartridge feeder guide knobs (e), and therefore inseparable from the sheet metal frame.

P-08 PARABELLUM

Tools And Their Use

The Cleaning Rod.

The cleaning rod is made of brass so as not to damage the bore and is used solely for cleaning the bore. Its handle forms a box with screwed lid for the grease (vaseline) to be used at the end of firing.

The Screwdriver.

Used for loosening or tightening the screws of the stocks, the only screws in the weapon, and if necessary, when removing or inserting the firing pin and the cartridge ejector. Furthermore, the handle part with the hole in it serves to help fill the magazine quickly and easily during continuous firing, being pressed down over the cartridge feeder guide knob so that the bend projects beneath on the right side and can be drawn off with the thumb when the cartridges are being inserted.

The Drive Punch.

The drive punch is used to remove the extractor pin, the automatic safety sear spring and the safety catch pin.

P-08 PARABELLUM

Final Remarks

The automatic pistol "Parabellum", since 1900 used all over the world in great numbers, has proved in practical use to be most reliable in every respect as both ordnance and a sporting arm.

That this pistol perfectly answers *all* the requirements is proved sufficiently by the fact that after very careful trials made for years comparatively together with different other known types and after practical military service trials in Germany, Switzerland, Bulgaria and Portugal, it has been accepted as an ordnance small arm. In other countries such as the USA, Brazil, Chile, Holland and Russia, large numbers have been bought for the troops.

P-08 PARABELLUM

Dimensions And Efficiency

	7.65mm	9mm
THE WEAPON		
Length of pistol (in center line)	9.25"	8.55"
Height	5.32"	5.32"
Distance between sights	8.38"	7.72"
Length of barrel	4.73"	4.02"
Rifling (right hand twist), 1 turn in:	9.84"	9.84"
Depth of rifling grooves	.005"	.005"
Width of rifling grooves	.122"	.098"
Number or rifling grooves	4	6
Weight of pistol	1 lb., 13.4oz	Same
Weight of magazine	1.93 oz.	1.93 oz.
THE AMMUNITION		
Length of loaded cartridge	1.173"	1.142"
Weight of loaded cartridge	162 grs.	193 grs.
Smokeless powder charge	5.4 grs.	5.4 grs.
Weight of ball (hard lead core with full or half case of sheet steel coppered & nickel plated	92.6 grs.	123.4 grs.

EFFICIENCY

		7.65mm	9mm
Velocity of projectile in fps		1150 (350 m)	1020 (310 m)
Working effect in foot pounds		271	283
Maximum range		1950 yds	1600 yds
with an angle of elevation equal to about		$27^{\circ}30'$	$30^{\circ}30'$
Penetration at 50 yards from muzzle			
in pine		6.3"	5.5"
in beech		2.76"	2.0"
Spread at 50 yds	height	3.9"	4.6"
	width	2.5"	3.2"
Spread at 100 yds	height	9.2"	11.0"
	width	6.4"	7.8"
Spread at 200 yards	height	24.0"	28.5"
	width	18.5"	23.0"

Frequency of aimless quick firing, magazines kept at hand, per minute of about 100 rounds. Time for reloading the pistol after the magazine is emptied, about 5 seconds.

Parabellums Of The World

The following is a comprehensive although not necessarily complete listing of various model Lugers. It is entirely possible, but unlikely, that other models also exist. Let's take a hypothetical but plausible example:

In 1910 a Chinese warlord orders 100 Lugers in 7.65mm, with 6" barrels and fixed sights. He also requests that the guns be marked with a dragon. The guns are made and delivered. Shortly before the outbreak of World War I (or during the war), a fire breaks out at the plant where they were made, destroying all the paperwork on them. Following the death of Sun Yat-Sen, there is no strong leader in the new Republic of China and fighting breaks out as various warlords engage in a struggle for power. Our hero's army is defeated by the army of another warlord near a river. The survivors are taken prisoner by the victors and are, of course, disarmed. The dead are stripped of their weapons. The confiscated weapons are loaded into a boat along with other supplies. While crossing the river, the boat sinks and all 100 Dragon Lugers are lost.

Presto! No records and no guns. Or, quite possibly, 2 or 3 of the victorious warlord's soldiers decided they liked their prisoners' Lugers better than their own heavy British Webley revolvers. They arm themselves with the captured pistols before the remainder are loaded onto the boat and carry them for the next few years, until the government is stabilized by Chiang Kai-shek in 1928. With China at peace, the soldiers "hang up their guns" in the best Western tradition. Less than 10 years later, the Japanese invade China and a Japanese soldier "liberates" one of the Lugers as a war souvenir. In 1942, a downed American fighter pilot (let's make him a Flying Tiger) is trying to escape from Japanese-held territory and return to his base. He is discovered by a small Japanese patrol and manages to kill them all with his trusty .45 in the best Sgt. York tradition. While searching through their belongings for food or any other supplies that might be of help, he finds and appropriates the Luger. He makes it back to his base, joins the U.S. Army Air Force when it takes over from the American Volunteer Group during the summer and flies safely until the war's end. He returns to the United States and brings the Luger along as a war souvenir. He is curious about the dragon marking, but assumes the gun was marked in China by a previous owner. In other words, a non-standard piece of no special value since the markings were neither factory nor government. He simply puts the Luger in a dresser drawer and forgets it.

If that same Luger were to turn up at a gun show today, and its history authenticated, it would be worth a small fortune. But with no factory records and no known pieces in circulation or in a collection, the Dragon Luger would, for all intents and purposes, simply not exist. This example is not really as far-fetched as it might seem, since only 4 Bolivian Lugers are known to collectors out of a much larger batch, and only 2 are known of an original group of 100 1903 Commercial Lugers. Many more of these two models probably do still exist, but are in the hands of people who have no idea of their value to a collector. Their *practical* value, of course, is no more than that of any other gun of comparable age, quality and condition.

Model	Barrel length	Caliber
1900 Commercial	4-3/4"	7.65mm
1900 Swiss Commercial	4-3/4"	7.65mm
1900 Geischert	4-3/4"	7.65mm
1900 Swiss Military	4-3/4"	7.65mm
1900 Eagle	4-3/4"	7.65mm
1900 Test Eagle	4-3/4"	7.65mm
1900 Prototype	4" & 7"	7.65mm, 9mm
1902 Test	4"	7.65mm, 9mm
1902 Commercial	4"	9mm
1902 Cartridge Counter	4"	9mm
1902 Eagle	4"	9mm
1902 Carbine	11-7/8"	7.65mm, 9mm
1903 Commercial	4"	7.65mm
1904 Navy	6"	9mm
1906 Commercial	4-3/4"	7.65mm
1906 Commercial	4"	9mm
1906 Swiss Commercial	4-3/4"	7.65mm
1906 French	4-3/4"	7.65mm
1906 Eagle	4-3/4"	7.65mm
1906 Eagle	4"	9mm
1906 Swiss Military	4-3/4"	7.65mm
1906 Swiss Police	4-3/4"	7.65mm
1906 Brazilian	4-3/4"	7.65mm
1906 Bulgarian	4-3/4", 4"	7.65mm
1906 Bulgarian	4"	9mm
1906 Dutch	4"	9mm
1906 Portuguese Army	4-3/4", 4"	7.65mm
1906 Portuguese Navy	4"	9mm
1906 Portuguese Navy R.P.	4"	9mm
1906 Russian	4"	9mm

1906 Navy Military	6”	9mm
1906 Navy Commercial	6”	9mm
1906 Chilean Contract	6”	9mm
1907 U.S. Army Test	5”	.45 ACP
1908 Commercial	4”	9mm
1908 DWM Commercial	4”	9mm
1908 Navy Military	6”	9mm
1908 Navy Commercial	6”	9mm
1908 Military (3 Issues)	4”	9mm
1908 Bolivian	4”	9mm
1908 Bulgarian	4”	9mm
1913 Commercial	4”	9mm
1914 Commercial	4”	9mm
1914 Military	4”	9mm
1914 Navy	6”	9mm
1914 Artillery	8”	9mm
1918 Spandau	4”	9mm
1920 Artillery	8”	9mm
1920 Commercial	3-5/8”, 8”	7.65mm & 9mm
1920 Spanish Guard	3-7/8”	7.65mm
1920 J. L. Galef	3-7/8”	7.65mm & 9mm
1920-22 Dated	4”	7.65mm & 9mm
1920 Swiss Commercial	4” & 4-3/4”	7.65mm & 9mm
1920 Swiss Rework	4” & 4-3/4”	7.65mm & 9mm
1920 Simson Rework, 1916-18	4”	9mm
1920 Abercrombie & Fitch	4-3/4”	7.65mm & 9mm
1920 Navy	4” & 6”	9mm
1920 Artillery	8”	9mm
1920 Carbine	11-7/8”	7.65mm & 9mm
1920 Carbine w/o Forestock	11-7/8”	7.65mm & 9mm
1920 Carbine w/o Square Grip	11-7/8”	9mm
1921 Krieghoff	4”	7.65mm & 9mm
1921 Krieghoff Squeeze Grip	3-7/8”	7.65mm
1923 Commercial	3-5/8”-20”	7.65mm & 9mm
1923 Commercial Safe/Loaded	3-5/8”-8”	7.65mm & 9mm
1923 Simson Military	4”	9mm
1923 Simson Commercial	4”	9mm
1923 Krieghoff Commercial	4”	9mm
1923 Stoeger	3-5/8”-16”	7.65mm & 9mm

1923(?) Pacific Arms	3-7/8"-8"	7.65mm & 9mm
1923 Stoeger Artillery	8"	7.65mm
1923 Dutch	4"	9mm
1923 Russian	3-7/8"	7.65mm
1923 Arabic Police	3-7/8"	7.65mm
Vickers Military	4"	9mm
Vickers Commercial	4"	9mm
Finnish Prison Guard	4"	7.65mm & 9mm
1924 Bern	4"	7.65mm & 9mm
1924-27 Simson	4"	9mm
1929 Bern	4-3/4"	7.65mm & 9mm
1929-33 Reworks	4"	9mm
1929-33 Riff Contract	4"	9mm
1930-33 Deathhead	4"	9mm
1933 Finnish Army	4"	7.65mm & 9mm
1933 Finnish Air Force	4"	9mm
1933 Stoeger	4"-11-7/8"	7.65mm & 9mm
1933 K.I.	4"	9mm
1933-35 Oberndorf Commercial	4"	9mm
1933-35 Oberndorf Dutch	4"	9mm
1934 Simson S Toggle	4"	9mm
1934 Mauser Prototype	4" & 4-3/4"	7.65mm & 9mm
Turkish Contract	4"	9mm
1934 Krieghoff Sideframe	6"	9mm
1934 Krieghoff Sideframe	4"	7.65mm & 9mm
1934 Krieghoff P Code Com.	4"	9mm
1935 Portuguese Banner	4-3/4"	7.65mm
Krieghoff S Code	4"	9mm
1936 Krieghoff	4"	9mm
1936 Persian	4" & 8"	9mm
1936-37 Dated Krieghoff	4"	9mm
1936-1940 Dutch Banner	4"	9mm
1936-1942 Dated Banner	4"	7.65mm & 9mm
1938 Krieghoff	4"	9mm
1939 Krieghoff	4"	9mm
1936-1939 S-42	4"	9mm
1937-1939 Banner Com. Dated	4"	7.65mm
1940 Krieghoff	4"	9mm
1940 Anshutz	4"	9mm

1941-1944 Dated Krieghoff	4"	9mm
1940 Mauser Banner	4"	7.65mm & 9mm
Mauser Banner Commercial	4"	9mm
Banner Commercial	4-3/4"	7.65mm
Austrian Banner	4"	9mm
Banner 2 Digit Dates, 41 & 42	4"	9mm
Lithuanian	4"	7.65mm
K Date S-42	4"	9mm
K Date S-42 Navy	4"	9mm
G Date S-42	4"	9mm
42 Coded 1939-1940	4"	9mm
442/1940 Deathhead	4"	9mm
S-42 Dated 1940-1941	4"	9mm
S-42 Nazi Navy	4"	9mm
42/41	4"	9mm
BYF Dated 1940, 1941, 1942	4"	9mm
C.C.C.P.	4"	9mm
1945 Dated Krieghoff	4"	9mm
Post War Krieghoff	4"	9mm
KU Marked	4"	9mm
Erma (now Stoeger)	4½" & 5½"	.22LR
1971 Mauser	4" & 6"	7.65mm & 9mm
1971 Mauser P-08	4" & 6"	7.65mm & 9mm

In addition to the above production models, there were, of course, a number of "one of a kind" Lugers such as Georg Luger's personal G.I. Baby Luger in 9mm with a 3" barrel and the Goering Presentation Luger. While interesting, these do not fit within the scope of this work since they were ego guns rather than working guns and it is doubtful if the presentation Krieghoffs were ever even fired by their owners. Also to be encountered occasionally are cutaways which were used for instructional purposes. These too are collector items but are worthy of mention since they did serve a practical purpose.

Tables I, II & III

TABLE I

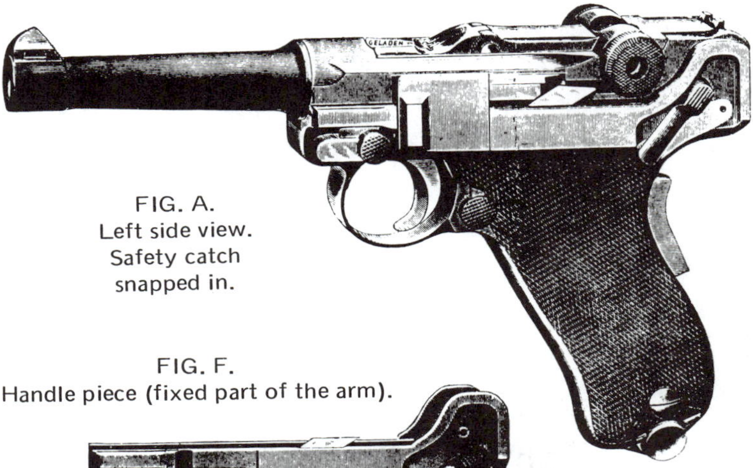

FIG. A.
Left side view.
Safety catch
snapped in.

FIG. F.
Handle piece (fixed part of the arm).

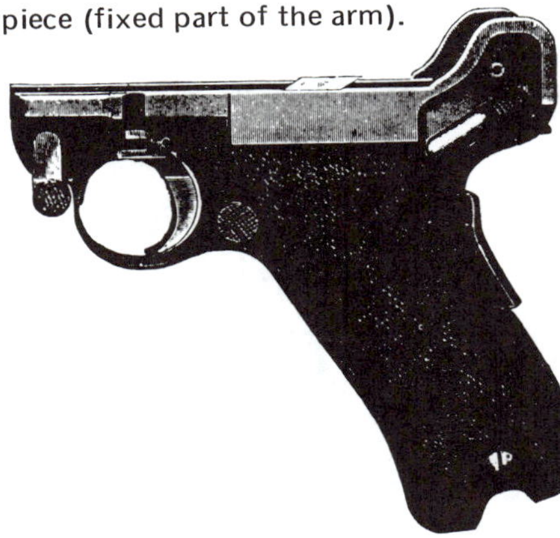

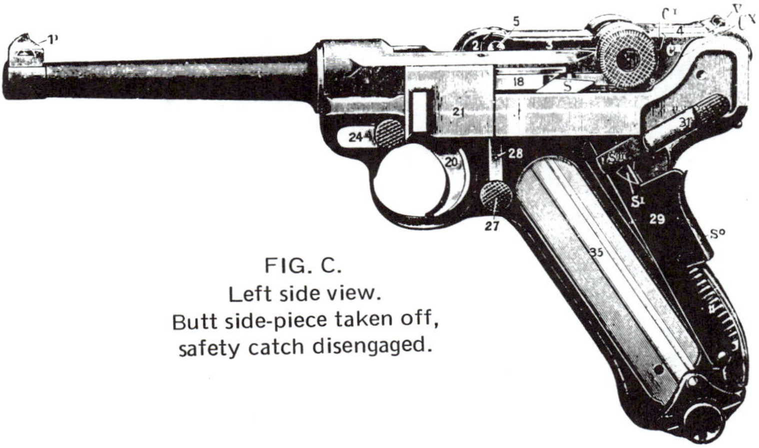

FIG. C.
Left side view.
Butt side-piece taken off,
safety catch disengaged.

TABLE I

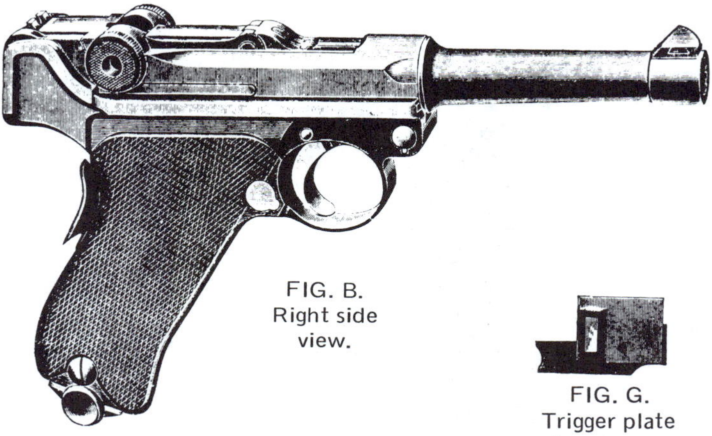

FIG. B.
Right side
view.

FIG. G.
Trigger plate
connecting the moveable
part with the fixed part.

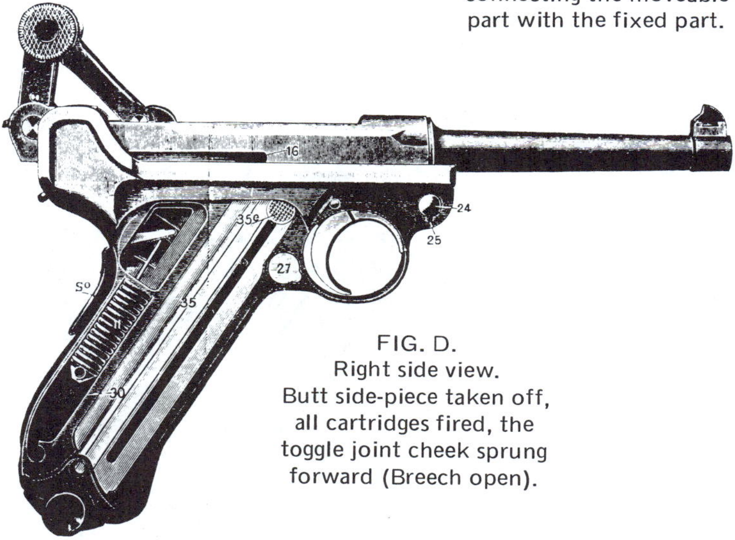

FIG. D.
Right side view.
Butt side-piece taken off,
all cartridges fired, the
toggle joint cheek sprung
forward (Breech open).

FIG. E.
Cannon with toggle-joint breech-block
(moveable part of the arm)

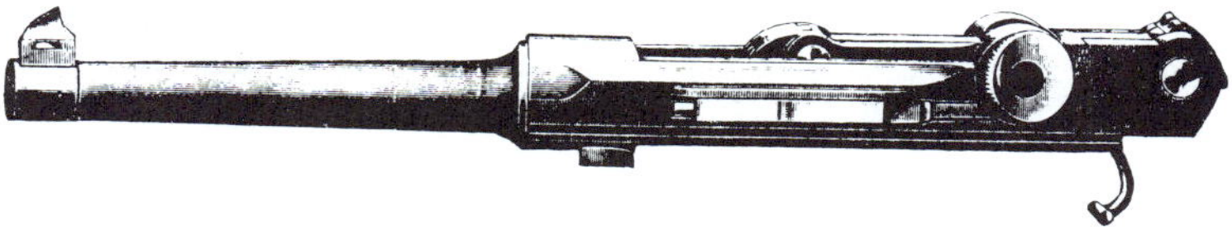

TABLE II

FIG. K.
Cross section before
the trigger.

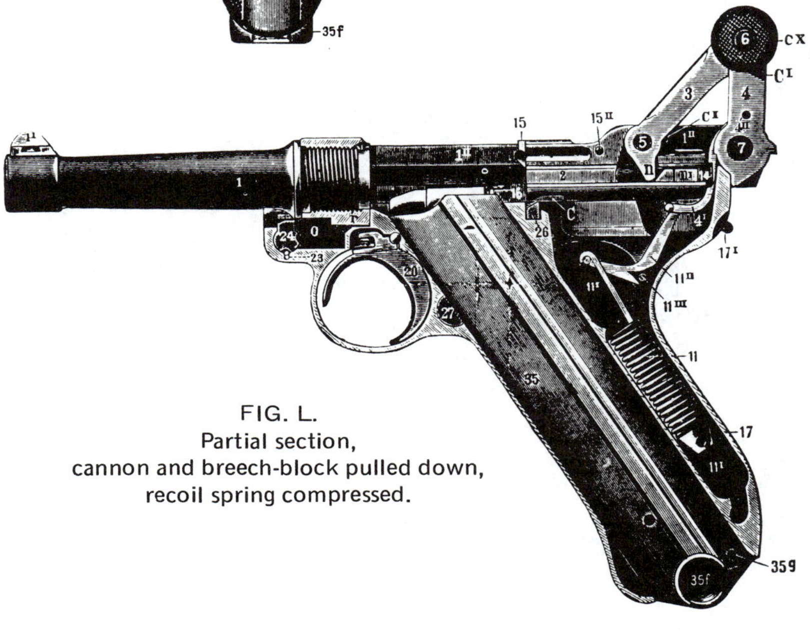

FIG. L.
Partial section,
cannon and breech-block pulled down,
recoil spring compressed.

TABLE II

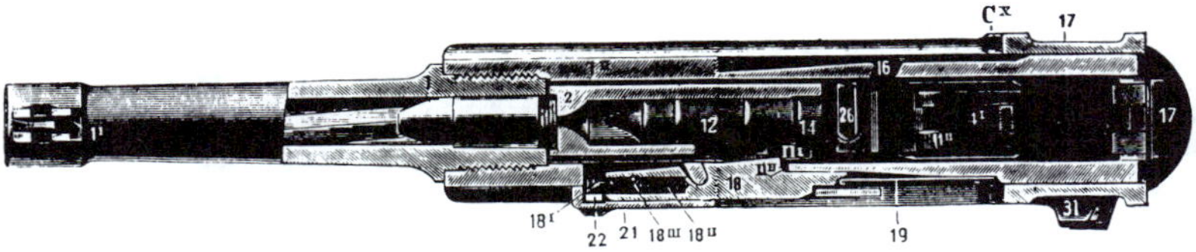

FIG. H.
Horizontal section.

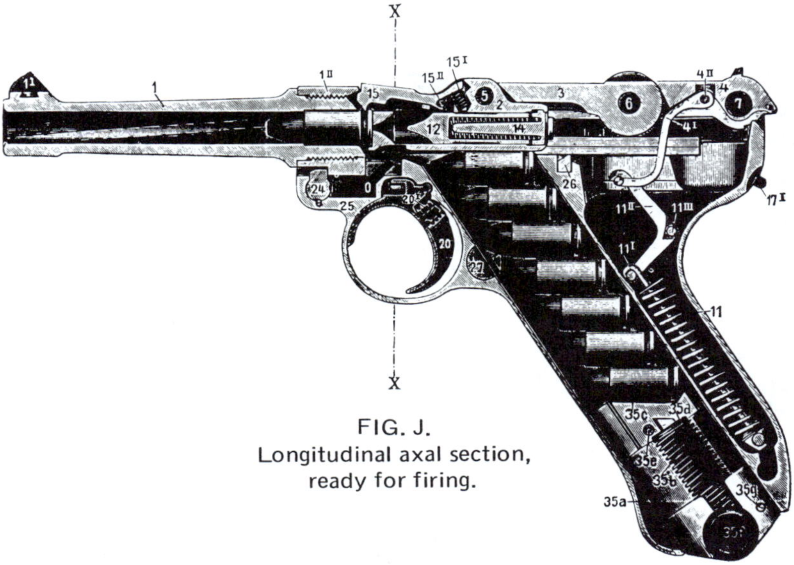

FIG. J.
Longitudinal axal section,
ready for firing.

FIG. M.
The immoveable frame, seen from above, with
the moveable parts of the breech-block removed.

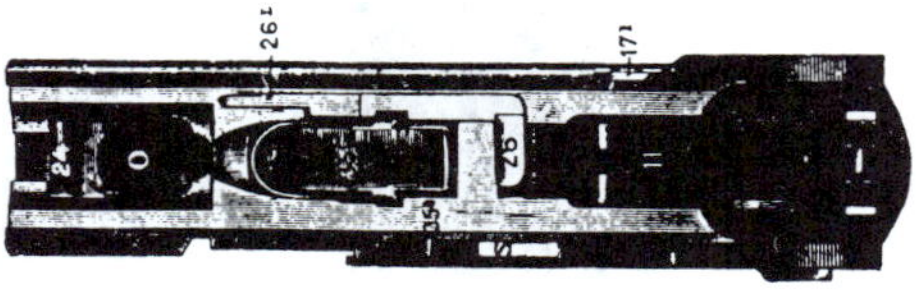

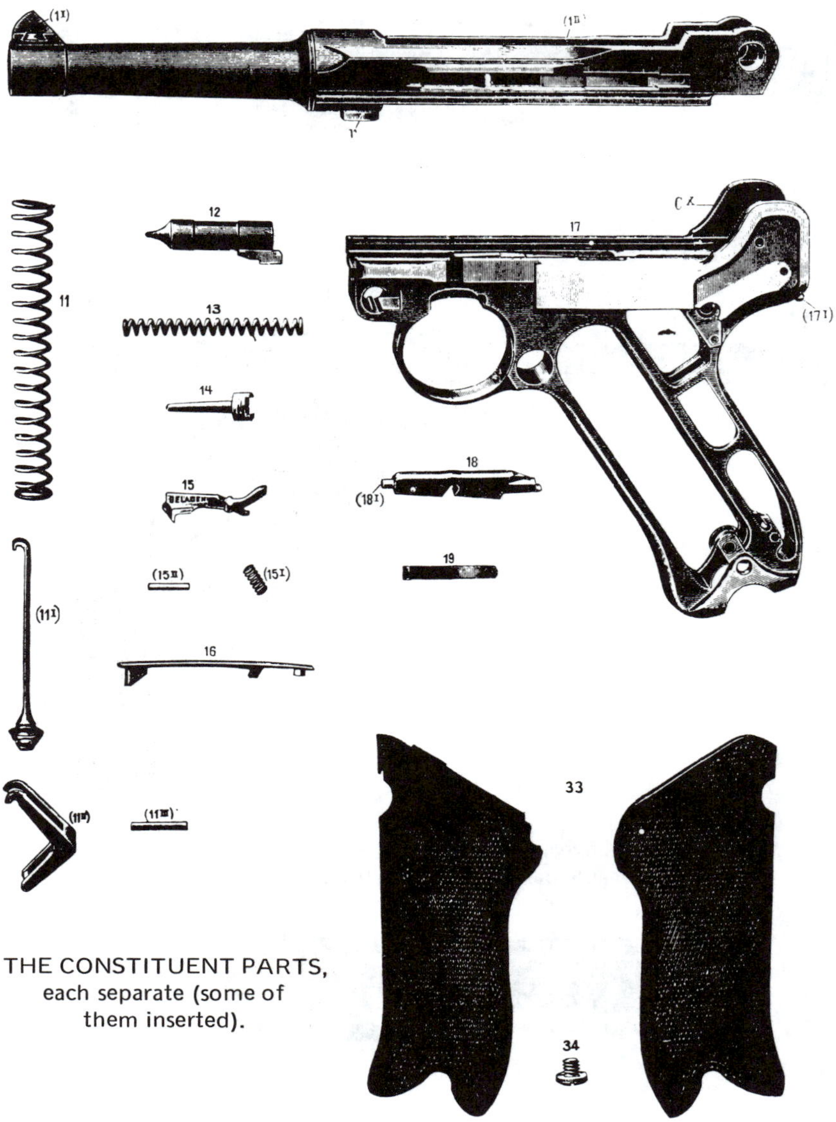

THE CONSTITUENT PARTS, each separate (some of them inserted).

TABLE III

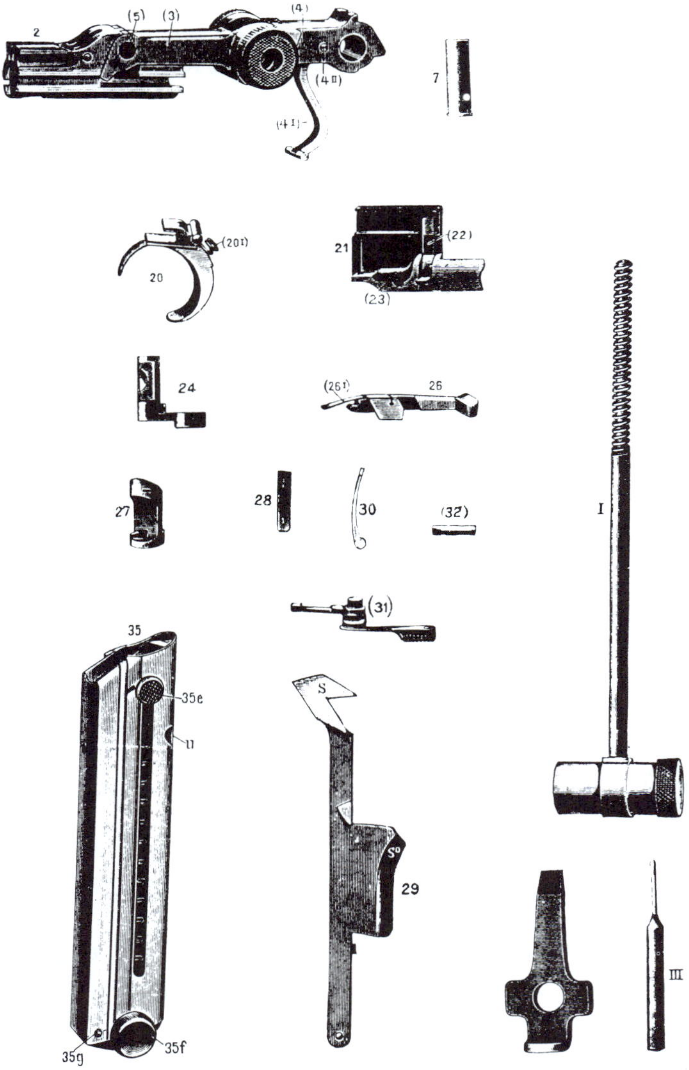

2
(5)
(3)
(4)
(4 II)
(4 I)
7
(20 I)
20
21
(22)
(23)
24
(26 I)
26
27
28
30
(32)
I
(31)
35
35e
11
35g
35f
S
S°
29
III

Notes

Notes

Notes

Notes

Other Books Available From Desert Publications

081	Crossbows/ From 35 Years With the Weapon	$15.95
206	CIA Field Expedient Preparation of Black Powder	$19.95
221	Evaluation of Improvised Shaped Charges	$19.95
411	Homeopathic First Aid	$14.95
454	Survival Childbirth	$15.95
C-002	How To Open A Swiss Bank Account	$9.95
C-011	Defending Your Retreat	$9.95
C-020	Methods of Long Term Storage	$9.95
C-023	Federal Firearms Laws	$11.95
C-024	Select Fire Uzi Modification Manual	$13.95
C-025	How to Build Silencers	$6.95
C-027	Firearms Silencers Volume 1	$13.95
C-029	M1 Carbine Arsenal History	$7.95
C-032	The Silencer Cookbook	$9.95
C-038	How To Build A Beer Can Morter	$4.95
C-040	Criminal Use of False ID	$14.95
C-045	Improvised Weapons of American Underground	$14.95
C-048	Catalog of Military Suppliers	$18.95
C-056	Navy Seal Manual	$29.95
C-065	Poor Man's James Bond Vol 1	$34.95
C-113	Submachine Gun Designers Handbook	$19.95
C-114	AR-15, M16 and M16A1 5.56mm Rifles	$13.95
C-117	Ruger Carbine Cookbook	$13.95
C-120	Shootout II	$14.95
C-122	AK-47 Assault Rifle	$15.95
C-123	Combat Loads for Sniper Rifles	$18.95
C-125	Browning Hi-Power Pistols	$13.95
C-126	UZI Submachine Gun	$13.95
C-127	FullAuto Vol 1 Ar-15 Modification Manual	$11.95
C-128	Full Auto Vol 2 Uzi Mod Manual	$11.95
C-130	Full Auto Vol 4 Thompson SMG	$11.95
C-131	FullAuto Vol 5 M1 Carbine to M2	$11.95
C-135	M-14 Rifle, The	$12.95
C-136	Fighting Garand Manual	$13.95
C-137	Ranger Training & Operations	$19.95
C-138	Emergency War Surgery	$29.95
C-140	Emergency Medical Care/Disaster	$14.95
C-143	Survival Medicine	$15.95
C-150	Improvised Munitions Black Book Vol 1	$17.95
C-151	Improvised Munitions Black Book Vol 2	$17.95
C-152	Impro. Munitions Black Book Vol 3	$29.95
C-155	Fighting Back on the Job	$14.95
C-162	Colt .45 Auto Pistol Manual	$13.95
C-163	Survival Evasion & Escape	$19.95
C-165	USMC Sniping	$19.95
C-175	Beat the Box	$7.95
C-197	How to Make Disposable Silencers Vol. 1	$19.95
C-198	Expedient Hand Grenades	$21.95
C-199	U. S. Army Counterterrorism Training Manual	$19.95
C-203	Explosives and Propellants	$19.95
C-204	Improvised Munitions/Ammonium Nitrate	$19.95
C-205	Two Component High Explosives Mixtures	$19.95
C-207	CIA Improvised Sabotage Devices	$19.95
C-208	Spcial Forces Demolitions Training Handbook	$19.95
C-210	Guerilla Warfare	$17.95
C-211	FN-FAL Auto Rifles	$21.95
C-218	Professional Homemade Cherry Bomb	$15.95
C-219	Improvised Shaped Charges	$19.95
C-220	Improvised Explosives, Use In Detonation Devices	$19.95
C-227	Springfield '03 Rifle Manual	$15.95
C-229	Lock Picking Simplified	$11.95
C-230	How to Fit Keys by Impressioning	$9.95
C-231	Lockout -Techniques of Forced Entr	$15.95
C-235	How to Open Handcuffs Without Keys	$13.95
C-244	Camouflage	$14.95
C-274	M14 and M14A1Rifles and Rifle Marksmanship	$23.95
C-284	Thompson Submachine Guns	$19.95
C-289	M1 Carbine Owners Manual	$13.95
C-290	Clandestine Ops Man/Central America	$14.95
C-324	How to Make Disposable Silencers Vol. 2	$19.95
C-337	Sten MKII SMG Construction Manual	$24.95
C-356	Map Reading and Land Navigation	$19.95
C-365	Survival Gunsmithing	$14.95
C-372	Poor Man's James Bond Vol 2	$34.95
C-386	Self-Defense Requires No Apology	$15.95
C-414	Select Fire 10/22	$15.95
C-415	Construction Secret Hiding Places	$15.95
C-416	Brown's Alcohol Motor Fuel Cookbook	$24.95
C-417	Canteen Cup Cookery	$9.95
C-418	Improvised Batteries/Detonation Devices	$19.95
C-421	Walther P-38 Pistol Manual	$13.95
C-422	P-08 Parabellum Luger Auto Pistol	$14.95
C-423	Sten Submachine Gun, The	$11.95
C-427	How to Train a Guard Dog	$16.95
C-429	The Anarchist Handbook Vol. 1	$14.95
C-430	Beretta - 9MM M9	$15.95
C-432	Mercenary Operations Manual	$14.95
C-439	Improvised Lock Picks	$17.95
C-443	How to Disappear Completely & Never be Found	$19.95
C-455	Modern Day Ninjutsu	$17.95
C-512	FullAuto Vol 8 M14A1 & Mini 14	$11.95
C-518	L.A.W. Rocket System	$14.95
C-528	With British Snipers, To the Reich	$29.95
C-539	Dental Specialist	$59.95
C-558	Food. Fur and Survival	$17.95
C-561	The Squeaky Wheel	$17.95
C-562	Survival Chemist Book	$17.95
C-564	Medical Specialist	$59.95
C-571	U. S. Marine Corps Scout/Sniper Training Manual	$27.95
C-610	Aunt Bessie's Wood Stove Cookbook	$9.95
C-622	Defensive Shotgun	$14.95
C-625	MAC-10 Cookbook	$14.95
C-627	Brown's Lawsuit Cookbook	$26.95
C-628	Can You Survive	$16.95
C-633	Hand to Hand Combat	$11.95
C-634	USMC Hand to Hand Combat	$9.95
C-635	Hand to Hand Combat by D'Eliscue	$11.95
C-636	Prisons Bloody Iron	$16.95
C-639	Police Karate	$17.95
C-646	Science of Revolutionary Warfare	$16.95
C-668	Survival Shooting for Women	$14.95
C-680	Infantry Scouting, Patrol. & Sniping	$16.95
C-694	CIA Field Expedient Incendiary Manual	$19.95
C-761	How to Lose Your X-Wife Forever	$24.95
C-763	Vigilante Handbook	$19.95
C-775	Alcohol Distillers Handbook	$27.95
C-776	How to Build a Junkyard Still	$15.95
C-777	Engineer Explosives of WWI	$19.95
C-794	US Marine Corps Essential Subjects	$19.95
C-815	US Marine Bayonet Training	$11.95
C-818	The Poisoner's Handbook	$29.95
C-823	Cold Weather Survival	$14.95
C-829	The Anachist Handbook Vol. 2	$14.95
C-840	Freedom Road	$21.95
C-889	Napoleon's Maxims of War	$11.95
C-890	Company Officers HB of Ger. Army	$19.95
C-891	German Infantry Weapons Vol 1	$22.95
C-896	Op. Man. 7.62mm M24 Sniper Weapon	$13.95
C-899	US Army Bayonet Training	$11.95
C-9002	German MG-34 Machinegun Manual	$19.95
C-9012	Firearm Silencers Vol 3	$18.95
C-9013	HK Assault Rifle Systems	$27.95
C-9047	SKS Type of Carbines, The	$17.95
C-9048	Rough Riders, The	$24.95
C-9052	Private Weaponeer, The	$17.95
C-9057	Keys To Understanding Tubular Locks	$11.95
C-9060	The Anarchist Handbook Vol. 3	$14.95
C-9080	Leadership Hanbook of Small Unit Ops	$14.95
C-9081	Dirty Fighting	$13.95
C-9082	Lasers & Night Vision Devices	$29.95
C-9083	Water Survival Training	$8.95
C-9084	Assorted Nasties	$29.95
C-9085	Ruger P-85 Family of Handguns	$19.95
C-9101	Guide to Germ Warfare	$17.95
C-9102	CIA Field Expedient Method for Explosives Preparation	$19.95
C-9119	Surviving Global Slavery	$16.95
C-9120	Military Knife Fighting	$14.95
C-9127	H&R Reising Submachine Gun Manual	$19.95
C-9131	Live to Spend It	$29.95
C-9134	Military Ground Rappelling Techniques	$15.95
C-9135	Smith & Wesson Autos	$19.95
C-9137	Caching Techniques of U.S. Army Special Forces	$14.95
C-9138	USMC Battle Skills Training Manual	$49.95
C-9139	Survival Armory	$27.95
C-9156	Concealed Carry Made Easy	$17.95
C-9164	The L'il M-1, The .30 Cal. M-1 Carbine	$17.95
C-9170	Urban Combat	$27.95
C-9178	Apocalypse Tomorrow	$17.95
C-9198	MP40 Machinegun Manual	$16.95
C-9199	Clear Your Record & Own a Gun	$24.95
C-9200	Sig's Handguns Manual	$19.95
C-9210	Sniper Training	$24.95
C-9212	Poor Man's Sniper Rifle	$19.95
C-9221	The Official Makarov Pistol Manual	$14.95
C-9224	The Butane Lighter Hand Grenade	$14.95
C-9228	Unarmed Against the Knife	$14.95
C-9229	Black Book of Booby Traps	$29.95
C-9239	Militia Battle Manual	$22.95
C-9247	Black Book of Revenge	$15.95
C-9253	Bankruptcy, Credit and You	$18.95
C-9255	The Sicilian Blade	$15.95
C-9259	How to Build Practical Firearm Silencers	$18.95
C-9262	Glock's Handguns	$19.95
C-9263	Heckler and Koch's Handguns	$19.95
C-9264	The Poor Man's Ray Gun	$11.95
C-9265	The Poor Man's R. P. G.	$24.95
C-9269	David's Tool Kit	$27.95
C-9279	Credit Improvement & Protection Handbook	$19.95
C-9281	Build Your Own AR-15	$19.95
C-9282	How to Live Safely in a Dangerous World	$19.95
C-9293	Bllack Book of Arson	$24.95
C-9322	How to Build Flash/Stun Grenades	$19.95
C-9342	Entrapment	$19.95
C-9359	The FN-FAL Rifle Et Al.	$21.95
C-9367	The Mental Edge, Revised	$17.95
C-9373	Survival Bible by Duncan Long	$59.95
C-9405	The Internet Predator	$17.95
C-9406	The Criminal Defendants Bible	$49.95
C-9430	Protect Your Assets	$24.95
C-9432	How To Build Military Grade Suppressors	$21.95
C-9441	Mossberg Shotguns	$24.95
C-9478	Poor Man's Primer Manual, The	$24.95
C-9546	Agents HB of Black Bag Ops	$14.95
C-9556	Improvised Rocket Motors	$19.95
C-9612	Homestead Carpentry	$14.95
C-9633	Getaway	$13.95

PRICES SUBJECT TO CHANGE WITHOUT NOTICE

Desert Publications

870-862-3811

215 S. Washington Ave Dept. - BK - 126

El Dorado, AR 71730 USA

Shipping & Handling
1 item $7.95 - 2 or more $12.95